Modern Farming

A Complete Guide to Farming Time Table & Practice

PROF. APIADE ADENUGA

INTRODUCTION

Fruit crops are labour and capital intensive and require technical knowledge to achieve the production potential. Effort has been made through these package of practices to make available the required knowledge and information for the benefit of fruit and crop growers. Our farming methods would be more efficient if we applied established principles that are well adapted to the agricultural production, the conditions and situation of the area producing such food. In the foregoing guide, we try to share some knowledge regarding best farming practices in order to direct food production durable farming, observant of the environment, producing good quality food conducive to good health while protecting labor conditions of small farmers and their families.

The main objective of the guide is to help the growing population of young farmers about the importance of implementing knowledge and best practices to improve the production system to yield more food so the country could be self sufficient in matters of nourishment. The ideas set forth in this guide will help the farmer produce better while taking the best action not only to counter climate changes but also to guarantee human health. This guide of Best Farming Practices (BFP) will help agricultural technicians, producers' organizations, school authorities, NGOs and all other instances concerned with farming.

TABLE OF CONTENT

1. Introduction	II
Caution	IV
2. Plantain Farming	1
3. Groundnut Farming	3
4. Cucumber Farming	6
5. Pepper Farming	8
6. Avocado	10
7. Pumpkin Farming	12
8. Cassava Farming	14
9. Yam Farming	17
10. Cabbage Farming	19
11. Corn Farming	21
12. Rice Farming	23
13. Beans Farming	26
14. Melon Farming	28
15. Potato Farming	30
16. Banana plantation	32
17. Pear Farming	34
18 Soyabean Farming	36
19. Scentleaf Farming	38
20. Beetroot	40
24. Management of other crop & horticultural pests	43

APPENDIX I - IV

I Some Other First Aid Measures	46
II Waiting Periods of Different Pesticides in Fruit and Vegetables Crops	49
III Fertilizer Sources for the Supply of Nitrogen, Phosphorus and Potassium	50
IV Some Other First Aid Measures	52

CAUTION

Chemicals used to control insects, diseases and weeds are poisons for human beings. Farmers are cautioned to use these poisons carefully to avoid any effect on human health. For safe use of these chemicals see Appendix III given at the end of this book.

Note :

1. For proper presentation of information on pesticides, fungicides, etc., it is sometimes necessary to use the trade name of the product or equipment. No endorsement of the named product or equipment is intended nor criticism implied of a similar product or equipment not mentioned in this book.

2. Volume of spray material to be used for controlling different insects and diseases of various crops is based on the usage of shoulder-mounted knapsack sprayer having "fixed type hollow cone nozzle." Spray volume may vary when other types of sprayers/nozzles are used for this purpose.

3. It should, however, be ensured that the actual amount of insecticides recommended in the "Package of Practices" should not be reduced. For proper control of weeds, it is always necessary to use flood jet or flat fan spray nozzles.

Plantain Farming

Plantain: Plantain cultivation is one of the simplest and most rewarding farming activities in Nigeria. The crop grows best in moisture-rich tropical climates. The flowers develop into a bunch of plantain, which holds about five to 15 strands of the fruits. Plantains do not have a growing season and are, therefore, available year-round, especially on irrigated farmland. This makes them a reliable food source.

Various requirements which are necessary to ensure successful plantain farming include improved varieties of planting materials called suckers; soil fertility; planting spacing know-how; manure application, and most importantly, water application.

When is the best time to plant plantain?

The best time to plant plantain is the rainy season as there is abundance of water. Suckers are planted immediately after field preparation. The humidity level during the rainy season also positively impacts the planting of plantain. However, they should grow vigorously and without water stress during the first 3 to 4 months after planting. Therefore they should not be planted during the last months of the rainy season.

How long does plantain take to grow?

Plantains are grown from a super long 12-to-15 foot underground rhizome. The resulting plant has giant leaves (up to 9 feet long and 2 feet across!) wrapping around a central trunk. Flowering takes 7 – 10 months of mild temperature and yet another 3 – 6 months to fruit. You can start harvesting the plantain bunches from the plantain trees after 10 months. Some varieties may take a little more than 10 months to reach the harvest stage. Ripening of the plantains usually take 1 – 2 weeks after full maturity of the plantains.

How many plantain trees can you get per acre?

500 – 1200 plantain trees can be planted on one acre of land depending on the crop spacing you use. Drip irrigated farmlands can take more plantain trees than those without drip irrigation system.

Which fertilizer is best for plantain?

NPK fertilizers, potassium nitrate, calcium nitrate and chelates etc. are the best fertilizers for plantain. Plantain is a potassium demanding crop so it requires fertilizers rich in potassium. Poultry compost manure, cow dung and pig excreta could also be generously spread on the plantation to enrich the land and get a bumper harvest.

Plantains always need extra nutrient if you want to get the best quality and manure or ash from wood fires can help in achieving this goal. It is important to use loamy soil, as it contains very good organic matters.

How you can control weed and pest infestation.

Proper clearing of the land is needed to avoid weed infestation. For weed control, one can decide to use herbicides or make use of manual labour and it is to be done every six to eight weeks at the early stage. Weed infestation declines after five to six months due to shading.

On **pest or disease** control, there are various ways, including cultural practices, mechanical and physical control, biological control and chemical control. The crops are to be protected from diseases because they are very harmful to them. There are improved varieties which were bred for disease resistance and low sugar content.

Water application

Like rice and sugarcane, plantain is a water-loving crop, and dehydration of the crop induces stunting, poor fruiting and susceptibility to drought and wasting. Drip irrigation or trickle irrigation is the best form of irrigation for plantain farming and indeed most crops.

Water is very important during the dry season. But with drip irrigation, your plantain plants can get water all year round. Most plantain farmers in Nigeria cultivate this crop in the rainy season. Their plantain plants get only 5-6 months of rainfall. Imagine, your plantain gets water through irrigation all year round; the yield will be 50-100% more.

The yield can even be more than 100% if you practice fertigation.

Fertigation is the act of passing nutrients in form of soluble fertilizers to plants through the drip irrigation system.

Plantain

Groundnut Farming

Groundnut Farming (or peanut farming) is a very lucrative agro-business venture. It guarantees a large return on investment, especially with the application of the right techniques. Because of its numerous uses, groundnuts are in great demand and this opens an opportunity for interested investors to export the product. Groundnut can be cultivated in a wide range of environment. However, it grows well in arid and semi-arid regions, requiring an optimum temperature of about 30°C.

When is the best time to plant Groundnut?
The best time for planting groundnut is around May-June. The soil should be warm at the time of planting, so it is important to get the land ready before then. Groundnut sowing kicks off in the raining season during the onset of the rainfall. If you are using irrigation farming system to grow your groundnut, the best thing to do is to moisten the soil before planting the seeds. By irrigating before planting the seed, you increase the growth rate of the groundnut crop.

How long does groundnut take to mature?
It takes approximately 4-5 months for the crop to mature. Some do mature as early as 90 days while others takes up to 130 days, depending also on climate condition.

Which is best soil for growing groundnut?
When choosing a site for planting groundnut, you must have in mind that groundnut produce their pods underground. Groundnut cannot stand frost, long and severe drought or water stagnation. The crop does best on sandy loam soils rich in organic matter with good drainage. Heavy and stiff clays are unsuitable for groundnut cultivation as the pod development is hampered in these soils. Thus the topsoil must have low clay content (less than 20%) with a loose structure. This will make it easy for the reproductive roots to penetrate the soil freely. The optimum germination temperature of the environment should be between 20-30°C with a minimum of 18°C. The temperature of the water absorbed by the seed is a critical factor for healthy germination. If the water temperature is initially low and gradually increases, we find reduced germination. Planting the seed in warm soil results in fast germination and healthy seedlings.

When can you harvest groundnut seed?

Harvesting usually consists of a series of operations comprising digging, lifting, windrowing, stocking, and threshing. You can know when your groundnut plant has reached maturity by observing the leaves. When you start seeing yellow spots spread across the growing peanut plant this indicates that your groundnut is ready for harvesting. You can also determine the maturity stage by checking the pods of the plant.

To do this, harvest random samples of the plant and cut out some groundnut pods from the root. Break the groundnut open to see inside the pod. The peanuts should nearly fill the pods. You must ensure that you harvest the groundnut before it passes the stage of full maturity; otherwise, an appreciable number of pods could be lost.

Spacing and Planting Depth for groundnut

Plant your peanut seeds on a ridge or on the flat ground depending on the nature of the soil. The exact spacing distance between each peanut plant will depend on the variety of the seed you are planting. If you are planting the variety with small seeds, the spacing should be at 50cm between rows and about 30cm between plants. Large-seeded groundnut types are spaced at 75cm between rows and 45cm between plants. It is advisable to give good space for each planted seed to aid good yield.

Planting depth must be 5-7.5 cm to ensure that the plant develops and produces optimally. A seed which germinates slowly as a result of deep planting takes longer to emerge and a substandard plant will be produced.

Shallow planting of seed (less than 50mm) can only be considered when enough moisture is available and the climate is moist. In situations where moisture is not limiting, 50mm is the ideal planting depth.

Nutrient requirement

Groundnuts are adapted to a soil with a pH level of 5.5 – 7.0. At pH levels out of this range, certain elements become unavailable e.g. iron and zinc. Being a leguminous crop, groundnuts can fix atmospheric nitrogen (N) with the aid of root bacteria thus; it does not depend on nitrogen fertilization.

Peanut plant requires sufficient levels of potassium (K) and phosphorus (P) for normal growth and development. Groundnuts prefer residual phosphorus (P) to freshly applied phosphorus. However, in fields where the level of phosphorus is low, it should be applied. An oversupply of potassium in the soil can induce a calcium deficiency.

Weed control in peanut farming

Weeds compete with the crop for moisture, nutrition, light, and space. Effective weed control implies good control of weed throughout the growing season. You can control weeds by using chemicals or machines and perhaps, a combination of the two. The ultimate choice depends on the species of weeds involved, the level of infestation and the size of your bud.

Groundnut

Cucumber Farming

Cucumber belongs to the Cucurbitaceous family and is an important summer crop cultivated throughout India. Cucumber plant has a climbing or trailing habit. The tender cucumber can be eaten raw or with salt in a salad. Cucumber is also used in daily cooking. Cucumber seeds can be used in oil extraction. The cucumber crop requires a moderately warm temperature and grows best at a temperature between 20°C and 24°C. It's not suitable for frost conditions.

What type of soil is best suitable for Growing Cucumber?

Cucumber can be grown in wide varieties of soil. However, there are some things to consider about the soil. Ideally, the soil should be loamy or clay and the land should have adequate access to water. The pH of the soil should be fairly neutral or slightly alkaline (close to 7.0). Additionally, the area should have sufficient exposure to sunlight since light is very essential for cucumber growth.

When is the best time for planting cucumber?

Cucumber can be planted at any time of the year. The plants generally need adequate water to thrive, so if you're planting during the dry season, there should be an irrigation system in place to ensure sufficient water supply. In mild climates with long growing seasons, plant them outdoors between April and June. In very warm climate, plant them as early as February or march through July.

The best season Seed rate and Seed Treatment of Cucumber

Sow the cucumber seeds during February or march through July.. About 3 kg of cucumber seeds is required for a hectare. Treat the cucumber seeds with Pseudomonas fluorescence 10 grams/kg or & Trichoderma viride 4 grams/kg or Carbendazim 2 grams/kg of seeds before sowing.

Where is the best place to plant Cucumber?

Cucumber thrives well in warm-humid weather, loose organic soil, and plenty of sunlight. They grow well in most areas of the United States and do especially well in southern regions. When planting cucumber, choose a site that has adequate drainage and fertile soil. You can plant on bed and you should ensure the right spacing between the plants. Also, you can plant your seed in the nursery then transplant later to the farmland. However, the recommended method is planting directly on the farmland due to the fragile nature of the cucumber root hence transplanting could damage these roots.

What month are cucumbers ready for harvest?

Cucumbers are generally ripe and ready for harvest anywhere from 50 to 70 days after planting. . However the time taken to mature depends on the variety of cucumber seed planted. A cucumber is usually considered ripe when it is bright medium to dark green and firm. You should avoid cucumber harvesting when cucumbers are yellow, puffy, have sunken areas or wrinkled tips.

CUCUMBER

Pepper Farming

Pepper is a perennial plant, but growers, in most cases, treat it as an annual. There are about four common varieties of pepper in Nigeria – Chilli (Ata ijosi), Habanero (Atarodo), Sweet pepper (Tatashe) and cayenne (Shombo). The highest in demand among the four is sweet pepper because of its uniqueness in taste and usefulness for meal decoration. It is not peppery which makes most people prefer it. However, all these varieties are needed for its specific culinary functions. It is crucial to decide the growing method as well as the varieties of pepper that thrive in your area. There are two methods to grow peppers: Growing from seed or growing from seedlings.

When is the best time to plant pepper?

Most commercial sweet pepper or chilies growers start the crop from seeds (hybrids) in an indoor protected environment. This take about 2-3 weeks before transplanting. As they wait for the young seedlings to grow and be ready for transplanting, they prepare the soil. Most pepper seeds sprout in about a week at a temperature of 70-80 degree °F, but germination can be spotty depending on variety.

How long does it take for pepper to reach maturity?

Peppers need a fairly long growing season to mature. Depending on the variety of pepper planted, they need 2-3 months from transplanting to harvesting and can take 4-5 months (100-150 days) to grow peppers from seed to harvest. Some may grow much faster than others.

Why do most pepper seed not germinate?

The restrictive factor when growing peppers outdoors is always the temperature. The optimum temperature is 18-26 °C (64.4-78.8 °F). The plant needs day temperature close to 23 °C (73.4 °F) and night temperature close to 18 °C (64.4 °F) in order to produce pollen. Soil temperature should not fall below 18 °C (64.4 °F). Cold weather during the growth period will inhibit the plant's growth. Shocked plants cannot recover easily.

Soil Requirements for planting pepper?

Peppers do not have strict soil requirements. They grow well in a wide variety of soils. However, the plant thrives best in medium to sandy-loamy soil with proper aeration and drainage. It is a sensitive plant to both drought and water-soaked conditions. The soil pH levels usually range between 6 -7. It does not grow well on alkaline soil.

Disease and pest control

During the nursery stage and on the farmland, fungicide and pesticide should be sprayed regularly to prevent diseases and pests and ensure maximum yield.

Weeding and fertilizer application

Weeding can be done thrice before harvest. Fertilizer; NPK should be applied 2weeks after transplanting. The other application should be during the early stage of flowering.

Do peppers need pruning? – How to prune peppers

Pruning is an important procedure and provides several advantages. However, not all pepper varieties need pruning. For those who do, pruning enables farmers to control the vegetation. Moreover, pruning leaves extra space for aeration preventing fungal infections. Besides, harvesting becomes much easier. In general, un-pruned peppers have a tendency to produce numerous peripheral sprouts and leaves. The extra foliage makes it difficult for the producer to manage the plant. The necessary pruning procedure includes the removal of the peripheral stems. Producers keep only the 2-4 sprouts on the plant. This way, the plant has a more flexible and manageable shape. Many producers also perform thinning. They remove the stems that grow between the sprout and the foliage. It is crucial to try not to cut the stem too close to the vein. Instead, you can consider keeping a 4 cm (1.6 inches) distance to avoid infections. On the other hand, pruning and staking increase labor costs, which may be balanced if the harvesting period is long enough. In the market, we can find determinate varieties that offer satisfactory yields without pruning.

Bell Peppers

AVOCADO

Avocado is a tropical fruit plant that looks similar to pear fruit. Avocado pear, originates from Mexico and is grown in; Peru, Kenya, India, New Zealand, Brazil etc. These fruits are commonly called as "Butter Fruits" in India. It is called butter fruit because of its creamy nature. The botanical name for Avocado is Persia Americana. They grow up to 60ft and above, the fruits have different shape ranging from oval, spherical, egg shape and pear shape. These fruits are a good source of nutrients. It is rich in fatty acids but low in cholesterol. They are also consumed mostly for their medicinal values. An ever green tree that is not capital-intensive but produces good yield, it requires low care and management. Temperature is between 17°C-27°C and annual rainfall of 1000-1500 for good yield.

When is the best time to plant Avocado pear?
Avocado trees do best at moderately warm temperature at about (60F to 85F) with moderate humidity. They can tolerate temperatures, once established of around 28F to 32F with minimal damage. Avoid freezing temperatures. Start planting early February to April when the rains are just starting. The spacing can be 8m×8m, 10m×10m, 12m×12m. This will give you about 100 stands in a hectare but if you want to have up to 300 stands you can use 8m×4m, 10m×5m, 12×6m. But the choice of spacing depends on the variety. Avocado can be propagated through seed planting in the nursery and grafting. Matured seed from the fruit are used for direct sowing in nursery beds or in polyethylene bags. The seedlings should be transplanted to the main field after 6 months of planting in the nursery. It can also be grown in pots and containers. When transplanting, dig pit 90cm by 90cm and fill it with a mixture of top soil and manure in the ratio of 1:1.

Climatic and Soil Requirement for Avocado Fruit Farming
Avocado plants are grown in tropical and semi-tropical humid areas. These plants cannot tolerate hot dry winds and frost. They thrive best in true tropical to warmer parts of the temperate area. A location on the southern side of the home or in a dip or valley will ensure protection from winds. These fruits can be grown on a wide variety of soils except on poorly drained soils as these plants are very sensitive to water-logging. Avoid the saline soils as these plants do not tolerate saline conditions. The optimum soil pH should be between 5.0 and 7.0 for better growth and yield.

Irrigation in Avocado Farming

Avocado requires humid condition during important stages such as flowering and fruiting stages. It must be watered properly immediately after transplanting for quick establishment. Avocado pear requires plenty of water to grow otherwise they may die off. Adequate Irrigation is necessary during dry season or when rain is in shortage. For commercial avocado farming, drip irrigation is best suited. Intense heat and dryness can lead to falling of leaves, flowers, fruit and branches.

Manure and fertilizer application

Apply urea after every 3-4 months about 15-20 tonnes for a hectare. Avocado plants are heavy feeders; they need Nitrogen in good amount to grow robustly. The plant can grow with organic matter only. But For best yield and growth of the avocado, apply inorganic fertilizers along with organic manures.

Seedlings of avocado should be applied with N: P_2O_5: K_2O in a proportion of 1:1:1 whereas older plants can be applied in the ratio of 2:1:2. Generally, the soils having pH value above 7.0 will show the iron deficiency which can be corrected by adding iron chelate @ 30 to 35 grams/tree.

Weeding

Weeding should be done early either mechanically or manually. At the early stage you can carefully apply herbicides.

Disease/pest control

The common pests of avocado plant are mealy bug, mites, scales etc. While the common diseases of avocado plant are fruit spot, root rot, leaf rot etc. These diseases and pests may vary from each verity, soil type and moisture/climate conditions. Insecticides and pesticides should be applied to eradicate the pests. Resistant variety can also be Cultivate to fight against pest infestation. To control the root rot disease, apply metalaxyl by mixing with soil just before planting the seedlings. Contact nearest horticulture department or agriculture department for appropriate pest and disease control measures in an avocado fruit farming.

How long does avocado plant take to fully mature?

Growing an avocado plant (Persea Americana) from a pit is fun and educational and it usually takes only two to six weeks for the plant to sprout. Plants grown from seeds will be ready for fruit harvesting in 5 to 6 years after planting. There are two varieties of fruits in avocado farming, purple and green. To determine the maturity of the fruits, the color change should be observed.

Usually, purple variety will change the color from purple to maroon whereas in green variety, color changes from green to yellow. Avocado fruits will be ready for harvesting when the seed coat changes its color from yellow-white to dark brown color. These fruits only soften after harvest and it takes 5 to 10 days for ripening.

Avocado fruit

Pumpkin Farming

The pumpkin belongs to "Cucurbitaceae" family. It is a native plant of West Africa and is mostly grown and consumed in the southern part of Nigeria. India is the second largest country producing pumpkin in the world after China. Pumpkin is a very tender vegetable. The seed do not germinate in cold soil, and the seedlings are injured by frost.

Soil requirement and preparation for Pumpkin plantation:

To plant any crop, the soil needs to be prepared. The land must to be cleared and tilled either mechanically or manually, for roots penetration and proper germination of the seeds. It thrives well in a wide variety of soils but sandy loam soil with good organic matter is best suited for pumpkin farming. Soil with good drainage and the PH range of 6 to 7 is ideal for pumpkin cultivation.

Best season for growing Pumpkins:

Pumpkin can be grown during the time period of January to March & September to December. Sowing can also be started after the first few showers from May to June for the rain-fed crop.

How to sow Pumpkin seeds:

Seeds of the vegetable are grown into seedlings in a nursery and are transplanted into the growing fields. The pod bearing the seeds are harvested at the end of a cycle, the seeds are extracted and dried for a day or two. They should not be over-dried to preserve the viability of the seeds. The drying of the seeds makes them resistant to pests that attack seed beneath the soil. It also prevents them from rotting.

It is required to sow 4 or 5 seeds per hole. Any unhealthy or damaged plants should be destroyed after two weeks. The hole must be 60 cm diameter and 30 to 45 cm in depth and spacing of 4.5-meter x 2.0meter. Farmyard Manure (FYM) and fertilizers should be applied with the topsoil in the pit/hole.

Manures and Fertilizers of Pumpkin:

Apply Farm Yard Manure (FYM) at 20 to 25 ton/ha as basal dose along with a half dose of 35 kg of N and a full dose of 25 kg of P_2O_5 and K_2O (25 kg/ha). The remaining doses of 35 kg of N should be applied in two equal split doses at the time of plantation and at the time of full blooming.

Pests and Diseases in Pumpkin farming:

Most diseases that affect pumpkin are airborne and most of the insects are boring ones. An example is caterpillar. Insecticides should be used in recommended quantity for pest and insect control, strictly following instructions on usage. Other pest that affect pumpkin plantation includes; Epilachna beetle, fruit flies, and red pumpkin beetle. They can be controlled as in case of bitter gourd. Powdery mildew, downy mildew, and mosaic are the main diseases in pumpkin farming. pumpkin farm should be properly fenced to protect them from goats and other herbivores animals.

Maturity and harvesting of pumpkin

With adequate water, pumpkin germinates 10 days after planting. Farmers can start harvesting two to four weeks after planting or when the stems are long. Pumpkins will be ready for harvesting when the stems connecting the fruit to vine begin to shrivel up. They can also be harvested when they have attained a deep solid color and the rind has become hard. The pod is ripe for harvest when the tendrils are dried; sometimes they fall off on their own. Harvesting may be done with the use of the hands or knives by cutting the stem a little distance away from the bottom of the stem. Cutting should be at the Position of the nodes. With proper irrigation and adequate manure, the vegetable should thrive in abundance.

Pumpkin plantation

Cassava crop

Cassava is a perennial shrub. It comes from Latin America. It is highly appreciated in West Africa and particularly in Côte d'Ivoire. The usually grow to height of between 1.5-2m (though some may grow to 4m). They are mainly grown for their starchy and tuberous roots. They are also widely consumed food in sub-Saharan Africa. Cassava has been utililized for year for several purposes, either for human or animal consumption or for processing:

At the domestic level, a wide variety of foods are made using a base of cassava: attiéké, foutou, gari, toh, ragoût, tapioca (semolina), placali, atoukpou, konkonte (Ghana) / konkonde (Côte d'Ivoire), etc. These foods are rich in iron. Cassava leaves are a nutritious vegetable for households. The peels of cassava can be transformed into food for cattle, buffalo, pigs, and chickens.

At the industrial level, cassava is used to produce liquor, beer (e.g. in Ghana, Cameroon and Mozambique), flour, bread, pellets, textile finishing (as starch), glue, and other products. The starch from cassava tubers is used in the production of plywood, paper, and textiles, among other things. Cassava is used as a raw material for the manufacture of sweeteners, fructose, alcohol, and ethanol fuel.

What is the best month for planting cassava

Cassava is grown between May and August and harvested between November and April. They should be planted when the soil is quiet wet, after the beginning of the rainy season.

What is the required spacing for cassava planting?

The quality recommended for 1 ha is 60 bundles of cassava stem. The stem cutting should be 25cm long and should be planted at a spacing of 1m by 1m.

How to plant cassava

- Stake to better distribute cuttings in the field.
- Keep 1 metre between stakes and 1 metre between rows. The recommended planting density is 10,000 cuttings per hectare.
- Buy cuttings one week before the May to June planting period.

- Choose cuttings of cassava varieties that are resistant to disease, drought-tolerant, and productive. When you obtain cuttings, make sure to procure additional cuttings to replace cuttings that do not germinate or thrive.
- To obtain quality cuttings, contact a multiplier supervised by an agricultural extension organization.
- An area measuring half a hectare (5,000 m2) requires 5,000 cassava cuttings; however, make sure to have some additional cuttings to replace any plants which fail or die.
- When planted alongside plantain, the recommended density is 3,000 cuttings per hectare.
- Prepare cuttings the day before planting to ensure good germination*. To prepare, choose cuttings with 4-5 eyes and cut stems into 20-25 cm pieces. Ensure that cuttings are 7-8 months old and not injured or diseased.
- Ensure that the cuttings are taken from the central part of the stem.
- When planting, make a hole of 10 cm deep. Cassava cuttings should be spaced 25 to 30 cm from each stake. This aerates the soil and helps the cassava tubers thrive.
- Plant one cutting per hole at an angle of 45 degrees, with more than half of the cutting in the soil and the nodes/eyes facing upwards. Avoid planting cuttings upside down, as this hinders germination.
- Replace dead plants/cuttings one month after planting.

Fertilization Application
- Fertilize 60 days after planting in August-September to maximize yield. For details on recommended products, ask an agricultural extension organization for advice.
- Add 20 g or NPK 10-18-18 for each cutting—five full soda caps. This is equivalent to 100 kg or 2 bags for a half-hectare plantation.
- Spread the fertilizer in a 30 cm circle around each cassava plant, then ridge* (gather the soil around the plants), while covering the fertilizer with soil.
- Alternatively, use organic manure such as chicken droppings or any other manure at a rate of 10 tons per hectare.

Pest and disease control

Insecticides and fungicides are not only dangerous for human health, but also for the environment. To use plant protection products safely, it is important to observe a minimum number of rules, especially those approved by specialized agricultural advisory organizations.

- Control pests (including locusts) from May to October by planting healthy (disease-free) cuttings from the central parts of the stem.
- Treat cuttings that are lightly infested with green mites, mealy bugs, and other pests by soaking them in hot water for 5-10 minutes just before cutting.
- Remove contaminated stems during the phase when cassava is growing, after the roots are harvested, and from stored bundles.
- To deter termite attack, cover the cut ends of the cutting with a wet paste made of a mixture of soil and oil.
- Observe the field and the cassava plants regularly to detect possible attacks as soon as possible.
- Treat with a recommended insecticide in case of major attacks or if insect pest problems persist, while respecting the rules of pesticide use.
- Apply insecticides only in the early morning or evening when there is no wind. All insecticide treatments must be stopped one month before the beginning of the harvest.

Weeding

- Weed the field in March, June, September, and December.
- If using herbicides, apply the rate indicated on the herbicide label. Note that there is no selective * herbicide for cassava yet. Instead, there are non-selective * herbicides, which should also be used to clear the field of weeds before planting.
- As the weeds compete with cassava plants for water, sunlight, and nutrients, it is recommended to remove them from the field as early as the first month after planting, using a machete or a hoe. This prevents weeds from growing to knee height (30-40 cm.)

Harvest

Harvesting of cassava must be done from November-December of the first year. Harvesting of the cassava should be arranged at the time which is suitable for the variety and intended use. cassava is usually harvested from 10 months after planting the cuttings or from the 12th month if the crop is intended for local consumption. Harvesting can continue for 8-9 months, as cassava has a life cycle of 12-24 months.

- Harvest by cutting the stem at about knee height with a machete or pruning shears*, and then pulling it out of the ground. Detach the tubers from the stem without injuring them in order to preserve quality.

Cassava stems

Cassava root tuber

Yam crop

Yams are starchy staples in the form of large tubers produced by annual and perennial vines grown all across the world. Yam is an important food crop, especially in West Africa: total world production is about 19 million tonnes per annum, some 70 per cent of it grown in Nigeria. They are regarded as mainly a source of carbohydrate. Yam tubers consist of about 21% dietary fiber and are rich in carbohydrates, vitamin C and essential minerals. Yams are boiled, roasted, baked or fried.

When is the best month to plant yam?

Yam are planted at the beginning of the warm weather. In West and Central Africa yam are best planted between February and April when the rainy season is just beginning, depending on whether in humid forest or on the savanna. These are the best periods for land preparation.

What is the best method of planting yam?

Yams are best grown by tubers. tubers that have been stored will usually begin to sprout as the weather warms up. Plant the sprouted yams in to the soil at a depth of 15 cm. Place the new growth upwards. Spacing should be 50 cm apart. Prepare furrows, space between rows about 50 cm and above

How many months does it take yam to grow?

Yams typically take about **14 weeks** to mature. They require deep, loose, textured loamy soil that is rich in organic matter.

How many months does yam stay before harvesting?

Depending on the variety, yams are harvested **6 to 12 months after planting**. Tubers are ready to harvest when the vines die back in late autumn/ when the tops of plants start to go yellow and wither. Do not leave the ripe tubers too long in the ground, otherwise they become bitter and may rot. Excavate carefully to avoid damaging the skin as this may lead to rot and a decrease in market value. start a fair way back from the leaf stem. Tubers can be stored for many months in a cool, dry place. With some varieties, only one crop is harvested.

What is the life cycle of yam?

Yam is a perennial plant, although it is treated as an annual one. Growth takes place in three distinct phases, which develop over 12 months and repeat indefinitely during the life of the plant. For this reason, it is said that the yam is a perennial crop grown as an annual. Harvested tubers normally stay dormant (do not develop sprouts) for 30 to 120 days depending on environmental conditions, the date of harvest, and the species. This means that only one crop cycle is possible per year, possibly restricting supply.

Pest and diseases of yam tubers

The major pests that affect yams include insects such as leaf and tuber beetles, mealy bugs, and scales; parasitic nematodes; fungi causing anthracnose, leaf spot, leaf blight, and tuber rot; and viruses, especially the yam mosaic virus (YMV). They are treated using pesticides.

Yam tuber

Cabbage Plantation

Cabbage is a popular plant of the species Brassica Family and is used as a leafy green vegetable. The only part of the plant that is normally eaten is the leafy head; more precisely, the spherical cluster of immature leaves, excluding the partially unfolded outer leaves. Cabbage is an excellent source of vitamin C. It also contains significant amounts of glutamine, an amino acid that has anti-inflammatory properties.

What is the best Soil for Cabbage Production?

Cabbage is grown in varied types of soils ranging from sandy loam to clay. It requires a pH ranging from 5.5 to 6.5 for higher production. It also thrives best when the soil is rich in organic matter and good drainage.

What is the best time to plant cabbage?

Cabbage seeds planted at the end of the summer, in August or late September, should easily germinate in the warm soil and warmer temperatures of the ending growing season. Cabbage may also be planted later in the year, during the fall, as long as the soil temperature average around 50 degrees.

Field Preparation for Cabbage Plantation:

Cabbage is a heavy feeder; it quickly depletes the soil of nutrients and needs a steady supply of water and nutrients throughout its growth. The soil should be prepared in advance by mixing in aged manure and/or compost. Soil should also be well-drained to prevent the roots from spliting or rotting. The land is prepared by plowing it 3 to 4 times. The first plowing should be done by soil turning plow, and the bulky organic manures should be spread in the field.

Transplanting and spacing of Cabbage Plants:

Normally 120 grams of seeds are required for one acre. Apply 480 Kgs of dry manure into a seedling bed of 160 m², and then sow the seed on the seedbed. This should produce sufficient seedlings for one acre of field.

Transplant the seedlings at 4- 5 true leaves stage, about 25 days after sowing. Usually, they are spaced 45 cm apart in double rows of 45-60 cm apart on each bed of 90- 100 cm wide.

Fertilizers and Manures requirement for Cabbage Production

It is recommended to use urea instead of Ammonium Sulphate where the soil is relatively acidic. If the soil is deficient of boron, 5 –10 kg/ha of borax should be applied before land preparation. For basal fertilizer, manure should be applied into the rows before chemical fertilizer. Chemical Fertilizers: Fertilizer application varies with soil fertility. Basal application before transplanting: 25:50:60 NPK kg / acre. First top dressing should be done 10-15 days after transplanting: 25:50:60 NPK kg / acre. Second application 20 – 25 days after first top dressing: 25:00:00 NPK kg / acre. Third application 10-15 days after second application: 25:00:00 NPK kg / acre. Boron & Molybdenum should be sprayed at the button stage.

Harvesting Techniques of Cabbage:

Cabbage is harvested when the heads reach full size and are firm.

Cabbage plantation

Corn Farming

Corn, (*Zea mays*), also known as **Indian corn** or **maize**, cereal plant of the grass family (Poaceae) and its edible grain. Corn is a tender annual and a member of the grass family that can grow from 4 to 12 feet tall. They are classified commercially; based mainly on kernel texture, include dent corn, flint corn, flour corn, sweet corn, and popcorn. Corn is used as livestock feed, as human food, as bio-fuel, and as raw material in industry.

When is the planting season of corn?

Planting date is crucial to the success of an organic production system. Planting too early results in slow growth and increases the amount of weed competition, the incidence of seedling diseases, and the likelihood of damage from seedling insects. On the other hand, planting too late results in a greater risk of drought stress, increased insect damage from second and third generations of european corn borers, and reduced yield from a decrease in intercepted sunlight due to decreasing hours of daylight. Nonetheless, maize crop sown early Feb and mid Jul for spring and fall seasons respectively produced better grain yield owing to higher RUE. In all locations, plant at least two days when average temperatures are above 65°F. Depending on the soil type, time soil preparation and planting date so that soils are moderately dry at planting to minimize the risk of seedling diseases.

How to Plant Corn

To speed germination, seeds should be moistened, then wrapped in moist paper towels, and stored in a plastic bag for 24 hours. The seeds should be sowed about 1½ to 2 inches deep and 2 to 4 inches apart in short, and also side-by-side rows to form a block, rather than one long row. Fertilize at planting time with a 10-10-10 NPK fertilizer, which will help the corn grow rapidly. Water must be applied at sufficient rate during planting time.

Pest and disease resistance

Resistance to common seedling, leaf, and stalk diseases is an important characteristic for hybrids in organic production systems. Some hybrids even tolerate insect pests such as european corn borer and southern cornstalk borer. Unfortunately, most hybrids do not have resistance to a wide range of diseases or pests. Growers should select hybrids that combine good early growth characteristics with good resistance to diseases that are major problems in their area. Cultural practices are very important for establishing a vigorous, full corn stand. Stand establishment can greatly influence pest populations as well as crop competitiveness and tolerance to pest feeding.

In fields where pests are historically abundant, do not plant organic corn if suitable, effective, and economical pest management options are not available.

Fertilizers and Manures requirement for Corn Production

Corn generally requires from 120 to 160 pounds of nitrogen per acre, 30 to 50 pounds of phosphorus per acre, 80 to 100 pounds of potassium per acre, and smaller amounts of sulfur and micronutrients to obtain optimum yield. Organic corn growers should design their systems so that the amount of nutrients added to the system offsets the amount removed in the grain or forage.

Harvesting

Early harvesting usually avoids crop damage from pests or hurricanes and prevents field losses resulting from ear drop and fungal pathogens. Probably the most important reason for timely harvest is the potential for yield reductions resulting from ear loss and ear rots due to stalk lodging, ear drops, and reductions in kernel weight. Corn can take from 60 to 100 days to reach harvest depending upon variety and the amount of heat during the growing season. Ideally, corn harvest should begin as soon as the grain reaches moisture levels of 25 percent or less. Under favorable conditions, corn should be ready to harvest in 10 days or less following the black layer formation at the base of the kernels.

Corn(maize)

Rice Farming

Rice is the third-largest crop production, after sugarcane and maize. The main producers of rice are the nations of China, India, Indonesia, Bangladesh, and Vietnam. Rice is a staple crop. More than half the people in the world, about 3.5 billion people, rely on its production. Not only is rice a key source of food but it is also good source of income for many smallholder farmers. Choosing the best planting practice depends on location and rice ecosystem, soil type, and the availability of input and physical labor.

Seed preparation

Rice has over 40,000 varieties and hybrids. It's important for a farmer to use healthy seed of locally adapted varieties to get a crop with a good potential yield. Growth management starts with proper planting or sowing practices. Rice crops can be seeded directly into the field, or sown in seedbeds and then transplanted in the field.

Site selection

In Northern Nigeria, rice can be grown in two main areas:

Lowland areas: These are lowlands on the edges of flooded fadamas (inland valleys) and irrigation schemes where water is available for 4½ to 5 months.

Upland areas: These are areas with good soil and rainfall of over 700mm. Select fertile land.

Seed rate: It is advisable to use good quality seed from a reliable source for sowing. If farmers intent to use their own seed, it is important to first sort out unfilled grains before sowing to enhance good germination.

Lowland rice: Use 50–60 kg/ha of seed.

Upland rice: About 40–50 kg of seed are required to plant a hectare when sowing is conducted by dibbling.

What season is best for growing rice?

The recommended time for sowing rice in Northern Nigeria is from April to June. Rice is generally grown as a wetland crop in fields flooded to supply water during the growing season. The actual timing of sowing should, however, be adjusted in accordance with the time of the establishment of the rains.

How long does rice take to grow?

It takes rice plants 4 to 5 months to reach maturity. The rice grows rapidly, ultimately reaching a height of three feet. By September the grain heads are mature and ready to be harvested. On average, each acre will yield more than 8000 pounds of rice.

Spacing and Depth of planting: Sow rice seeds by drilling in rows at spacing of 20 cm or 30 cm apart. For transplanting method, transplant seedlings at a rate of 2–3 seedlings per hill, to a depth of 3–4 cm, and at a spacing of 30 × 30 cm (best for late-maturing cultivars), or 20 × 20 cm when soil is fertile or sufficient fertilizer is available.

Plant the rice seed at a depth of 2 to 4 cm. When rice is planted at a depth of more than 5 cm, the emergence of the young seedlings is delayed.

Fertilizers

The amount of fertilizer to apply depends on the quantities and level of residual nutrients in the soil and the type of fertilizer materials available. Soil fertilization can be practiced with mineral and organic fertilizers, depending upon the production type required. However, organic fertilizers are always recommended practice for the improvement of soil organic matter. The water level on the field at the time of fertilizer application must be maintained at 3–5 cm to ensure the efficient use of the applied fertilizer. If the water on the field is more than 10 cm, it will cause a loss of nitrogen fertilizer through volatilization, therefore, drain the field before fertilizer application.

Apply 60–80 kg of nitrogen and 13 kg of phosphorus (i.e., 30 kg P_2O_5/ha) and 25 kg potassium/ha (i.e., 30 kg K_2O/ha). The nitrogen should be applied in two doses in between stands properly incorporated (buried) in the soil to avoid losses. This is about 4 bags of NPK 15: 15:15 applied at sowing and about 2 to 3 bags of urea applied at 6–8 weeks after sowing.

Weed control

A number of methods can be combined to control weeds in rice. Examples of such methods are fallowing, land preparation, use of a competitive rice variety, water control, hand weeding, and herbicides.

Pest and disease control

Rice is less affected by field and storage pests than other cereals grown in Northern Nigeria. The major pests of rice are borers and armyworms. The borer cuts off the growing part of the plants from the base, feeds inside the stem, and causes the plant or tiller to die. Upland rice is also commonly attacked by armyworms and termites

(root-feeder), especially when the rain stops at the beginning of the season.

Chemical control: Application of Cypermethrin or Lamdacyhalothrin at 1 L/ha is effective against borers and armyworms. Chemical control measures should be implemented when about 20% of the field is infested.

Cultural control: After harvest, burn all stubble from the previous crop. This will destroy the semi-active resting stages of the borers that normally inhabit the stubble. Alternatively, the infested rice field could be flooded after harvest for a week to completely submerge the rice stubble, thereby drowning the borers.

Harvesting

Rice is ready for harvesting when the grains are hard and are turning yellow/brown (about 30–45 days after flowering). Rice is fully mature for harvesting when 80–85% of the grains are straw color. To harvest, cut the rice stems with a sickle at about 10–15 cm above the ground. Tie the panicles in bundles. Then place the tied-up bundles of the harvested rice crop in an upright position for drying before threshing.

Rice farming

Beans Farming

A **bean** is the seed of one of several genera of the flowering plant family Fabaceae, which are used as vegetables for human or animal food. Beans are a summer crop that needs warm temperatures to grow. Beans are high in iron, potassium and magnesium, and are also an important source of protein and fiber.

When is the best time for planting beans?

Beans are best planted towards the end of the rainy season. Intense rainfall during the rainy season months can negatively impact the growth and yield of the beans plants. Beans farmers in Northern part of Nigeria usually grow beans or cowpea in August.

Soil requirement

Beans grow well in good soil that is fairly loose. This crop does not do well in compact soil. The best sites should have full sun (partial shade is tolerated but will reduce the yield). It requires a well-drained soil (but consistently moist), average fertility (too rich of a soil will produce an excess of foliage at the expense of beans), slightly acidic soil pH (6 to 6.8), and good air circulation.

Fertilizer requirement

Beans farm requires farm inputs like fertilizers, insecticides, fungicides and nematicides. Beans require fertilizers that have high phosphate content. Examples are Single Super Phosphate (SSP) fertiliser or Diammonium Phosphate (DAP).

A little bit of Muriate of Potash (MOP) should be applied at the flowering stage of growth. High nitrogenous fertilizer can negatively impact beans crop.

Irrigation requirement

Beans require about 1 inch of water a week for better growth. Drip irrigation is the best type of irrigation for the growing of beans. Beans can be grown all year round if adequate irrigation is maintained.

How long does a bean take to mature?

They mature in 50 to 55 days, while pole beans will take 50 to 60 days. On average, it takes beans plant a little less than two months to fully mature, but different cultivars have different maturities. They time can vary from 45 to 75 days, depending on the cultivar and climate.

Diseases and Pests of Beans

Diseases and pests are prevalent in beans or cowpea crop in Nigeria. Most of them includes: Grain Borer, Blight, Leaf Spot, Root Not Nematodes, Brown Rust, Rhinoceros Beetle, Whiteflies,

Bollworm, Aphids, Mites, Viral Diseases etc. various control measures has been put in place to prevent Mexican Bean Beetle & Bean Leaf Beetle and many other pest form attacking the plant. The Control measures include cultivating to destroy overwintering locations; handpicking eggs, larvae or adults; using floating row covers; planting a heavy bean crop in the spring (beetle populations are heaviest later in the summer); using a trap crop; using predators; spraying with azadirachtin, garlic, cedar oil, or mineral oil; and planting less-preferred types like mung beans, cowpea, and soybeans. This is one of the top insect pests in many areas.

Harvest

Beans varieties have a short growth time. Some variety are harvested during a 40 days period while many other variety may take up to 100 days to reach maturity stage. The creeping variety has a longer duration time and higher yield. The erect type has a short duration but a slightly lower yield. You should harvest when the pods are dry and slightly brownish in color.

Beans farming

MELON

The melon (egusi) plant which comes from cucurbit species (family Cucurbitaceae) is a vine with climbing and creeping habits. Its leaves are deeply lobed, blue-gray in colour, slightly scabrid, denticulate and about 24 centimetres long. Melon has many health benefits which make it an important staple in the average Nigerian diet. One of these benefits is its protein content which makes it an excellent nutrition source. In fact, 40 percent of the seed is made up of protein. Other components of melon seeds include palmitic, stearic, linoleic and oleic acids which are essential in protecting the heart. Aside being a good source of protein, oil is sometimes extracted from its seeds. This is used in cooking and is sometimes processed into salad oil. It is interspersed among other crops, a combination commonly appreciated because its plants take care of weeds.

Site selection

Melon plant grows well in loamy soil, rich in organic content. Soil must be free from debris of plant remains or materials that will encourage fungal growth. Avoid overcrowding of plants by maintaining the recommended row and intra-plant spacing for the species/varieties grown. Soil tests can help to determine if there is need to apply fertilizer and/or soil conditioners to assure adequate soil pH and plant nutrition to avoid plant stress. It is also best to plant the seeds in a place sheltered with trees, where these trees can serve as support for the climbing tendrils. In the absence of trees to support the climbing vine, use stakes to support the melon plant. It takes about 6 months for the plant to mature.

Cultivation

The planting season for melons is between April and June, which is the start of the rainy season. Melon grows very well in an arid climate. They are planted 2-3 seeds per hole, 1.5-2 cm (0.5-0.75 in) in depth, and 1 m (3 ft) apart in holes. It is recommended to plant in the month of May in Nigeria. Emergence occurs in 4-7 days. Flowering occurs four weeks after planting and vines form a nearly complete ground cover, suppressing weed growth. Effective ground cover using Egusi Melon for crop inter-planting may be achieved with 20,000 Egusi Melon plants per hectare. Egusi Melon tolerates dry to wet growing conditions but fruits mature only in dry conditions 4 5 months after sowing.

Pests and Diseases

Egusi Melon is relatively disease free in its native regions. Variegated locusts eat Egusi Melon seedlings. The beetles *Tribolium castaneum* and *Lasioderma serricone* can seriously damage stored seeds. Seeds should be stored in well sealed container to prevent beetle damage. Minimize insect damage and fungal infection by proper use of acceptable insecticides, fungicides and other appropriate practices within an integrated pest management program.

Harvesting

Melons require ample supply of moisture for growth and fruit setting. However, too much water will diminish their flavour. When harvesting melons, you must cut the melon from the vine instead of pulling. This is because pulling creates a cracking wound that pathogens can easily penetrate and quickly destroy the quality of the fruit, while also ruining its appearance. Melon gourds are usually harvested between October and December. While melon seeds can be stored for as long as possible, they are susceptible to fungal attacks from disease-causing strains of the Penicillum and Aspergillus species. Such attacks tend to reduce seed germination, increase seed discoloration and also reduce the nutritional benefits of the plant by producing toxic metabolites and aflatoxins. Melon should be harvested after they stop enlarging; they can be kept well for several months in storage. Seed removal from the solid pulp requires breaking the fruit with a hard stick (not a machete which will slice some of the seeds) and laying the pieces pulp side down on the soil for several days.

Melon

Potato Farming

Potato, (*Solanum tuberosum*), annual plant in the nightshade family (Solanaceae), grown for its starchy edible tubers. The potato is native to the Peruvian-Bolivian Andes and is one of the world's main food crops. The potato is the world's fourth-most important crop after rice, wheat and maize, and the first among non-grains. Potatoes are frequently served whole or mashed as a cooked vegetable and are also ground into potato flour, used in baking and as a thickener for sauces. The tubers are highly digestible and supply vitamin C, protein, thiamin, and niacin.

Soil requirement

Potatoes can be cultivated in a variety of soil types, including loamy soil, sandy loam, silt loam, and clay soil. Soil should be loose in order to provide less resistance to tuber enlargement. The soil must be fertile and well-drained. Potato cultivation requires acidic soil with a pH range of 4.8 to 5.5. The potato is classified as a cool-weather crop.

When is the best month to plant potato?

Potatoes are planted as early as 4 to 6 weeks before the average last rain in spring or any time after the soil temperature warms to 40°f. Potato is propagated by tubers. The tubers used should be medium in size (20-125 g weight) and bought from a reputable source.

How long do potatoes take to mature?

Potatoes can take up 70 to 120 days before maturity depending on the variety of potato planted.

Tuber Treatment

Potato seed/tuber must be kept in a cool, shady place for one to two weeks after being removed from cold storage to allow sprouts to emerge. Tubers should be treated with Gibberellic acid 1g/lit of water and then dried in a shady place before being kept in an aerated chamber for 10 days to achieve uniform sprouts. Tubers are soaked in a 0.5% Mancozeb solution for 10 minutes in order to prevent tuber rotting.

Space and depth of planting

The planting distance preferred for potato crop should be 50*20 cm and 60*25 cm, seed rate of 1.40 kg for 10 meter square of area. Potatoes are best grown in rows. Prior to planting, the soil must be cultivated. This will remove any weeds and will loosen the soil and allow the plants to become established more quickly. Potatoes should be planted one foot apart in a 4-inch deep trench, eye side up.

Manure and Fertilizer Requirement:
Apply 250-400 g/ha Farm Yard Manure, 120-160 kg/ha Nitrogen, 80-120 kg/ha Phosphorus and 80-120 kg/ha Potassium to the soil, 2-3 weeks before planting. At the time of sowing, apply $3/4^{th}$ dose of Nitrogen, and a full dose of Phosphorus and Potassium fertilizer. Apply also, $1/4^{th}$ dose of Nitrogen fertilizer at the time of earthling up operation, after 30-40 days of sowing. Earthing up operation is performed for proper aeration, temperature, and moisture maintenance in the soil. In this, the soil is drawn up around the base of the plant for proper tuber formation; it is done when the plant attains a height of 15-20 cm. The second earthing operation can be performed two weeks after the first one if required.

Irrigation Requirement
Potato crops require frequent irrigation, depending on soil moisture. The potato vines should be well watered throughout the summer, especially during the period when the plants are flowering and immediately following the flowering stage. During this flowering period, the plants are forming their tubers and steady water supply is crucial to good crop yield. Potatoes do well with 1-2 inches of water or rain per week. Irrigation must be turned off 10-12 days prior to harvesting. When the foliage turns yellow and begins to die off. This will help start curing the potatoes for harvest time.

Common Pests and treatment of Potato Plantation:
Potato crops are attacked by various pests and diseases. Most of these pests include Aphids, Flea Beetles, Leaf Hoppers, Early/Late Blight, and Potato Scab: Most likely caused by soil with high PH. Remember Potatoes like acidic soil (do not plant in soil with a pH higher than 5.2). Dust seed potatoes with sulfur before planting.

Treatment for pest infestation
1. **Aphid**- They causes curling and deformation of young leaves. Cut the foliage according region timing and spray Actara 25WG600g/ha.
2. **Leaf eating caterpillars**- They feed on potato leaves and damage them. Spray Profenofos @ 2ml.
3. **Cut Worms**- They cut sprout at ground level and damages the crop. Spray Tracer 48 SC 200 ml/ha.

4. **Potato tuber moth**- This is the most common pest in both the field and the storage. To protect the crop from this, use only healthy seeds and well-decomposed cow manure. It is a very effective approach to cover the soil with a 2-inch layer of sand.

5. **White grub**- They cause crop damage by feeding on the root, stem, and tuber. Collect larvae while tilling soil, planting, and weeding to manage this pest. Also, deploy bird predators to minimize the population of white grubs.

Diseases
1. **Early blight**- Necrotic spots are observed on lower leaves. Spray Chlorothalonil fungicide.
2. **Late blight**- Observed on the lower part of leaves and tip. Spray Chlorothalonil fungicide.
3. **Black scurf**- Tubers with black scurf. Affected plants wither and die. Seed treatment with mercury is required to defend against this. Also, implement crop rotation and avoid monoculture. Brassicol should be applied to the soil.
4. **Common Scab**- This disease spread rapidly in low moisture conditions. To protect crops from this, avoid deep planting of tubers, follow crop rotation.

Harvesting

Potatoes are ready for harvest when the leaves begin to turn yellow and fall off by themselves. Irrigation is withheld before harvesting and harvesting is done when soil becomes dry. Care should be taken when digging out the tubers to avoid injury. Harvesting is done with the help of kudali or potato digger or by plowing with deshi plow. Potatoes are laid on the ground and allowed to dry in the shade after harvesting. The typical yield of a potato crop ranges between 20 and 30 tonnes per hectare.

Potato farming

Banana Farming

Banana is the oldest and commonest fruit known to the mankind. It is one of the important fruits, and constitutes second largest fruit industry in India. It is nutritious palatable and easily digestible fruit. It is available throughout the year. Banana is rich in carbohydrates, minerals such as calcium, potassium, Mg, Na and phosphorous. Other than fresh fruits, it can be consumed as processed in various forms like chips, powder, flakes, etc. Banana pseudostem is chopped and used as cattle feed. Also, the leaves are used as plate. The botanical names of banana are *Musa cavendish* and *Musa paradisiaca*, which belongs to the family Musaceae.

Climate and soil requirement

Banana, basically a tropical crop, grows well in a temperature range of 15°C – 35°C with a relative humidity of 75-85%. In India, this crop is being cultivated in climates ranging from humid tropical to dry mild subtropics through a selection of appropriate varieties. Rainfall is most important for vigorous vegetative growth of bananas.

Banana performs well in a deep, rich **loamy** soil with a pH ranging from 6.5 – 7.5. The soil should have good drainage, adequate fertility, and moisture. Saline solid, calcareous soils are not suitable for banana cultivation. Banana farming requires soil that are neither acidic nor alkaline, rich in **organic** material with high nitrogen content, adequate phosphorus level, and plenty of potash is good for a banana.

Different Varieties of banana

Banana has more than 20 varieties which includes; Dwarf Cavendish, Robusta, Monthan, Poovan, Nendran, Red banana, Nyali, Safed Velchi, Basarai, Ardhapuri, Rasthali, Karpurvalli, Karthali, and Grandnaine etc. Among all, Grandnaine is the most preferred variety due to its tolerance to biotic stresses and good quality bunches. Bunches have well-spaced hands with a straight orientation of figures, bigger in size. Grandnaine develops an attractive uniform yellow color with better shelf life & quality than other cultivars.

Method of Planting Banana

About 70% of the farmers in India are using suckers as planting material while the rest 30% of the farmers are using tissue culture seedlings. Sword suckers with well-developed rhizome, conical or spherical in shape having actively growing conical bud and weighing approximately 450-700 gm are commonly used as propagating material. In Banana Farming, suckers generally may be infected with some pathogens and nematodes. Similarly, due to the variation in age and size of the sucker, the crop is not uniform, harvesting is prolonged and management becomes difficult. Therefore, in-vitro clonal propagation i.e. Tissue culture plants are recommended for planting.

They are healthy, disease-free, uniform in growth, and early-yielding. Traditionally banana growers plant the crop at 1.5m x 1.5m with high density. And then the suitable spacing of 1.82 m x 1.52 m is being recommended, it accommodates 1452 plants per acre (3630 plants per hectare) keeping row direction North-South with wide spacing of 1.82 m between the rows. . A pit size of 0.5 x 0.5 x 0.5 m. is normally required. Small pits are dug in case of ridges and furrows. The pits are to be refilled with topsoil mixed with 10 kg of FYM (well decomposed), 250 gm of neem cake, and 20 gm of carbofuran. Prepared pits are left open for 15-20 days for solar radiation to kill all the insects, soil-borne diseases, and for aeration before refilling.

Irrigation Management of Banana Orchard:

Banana, a water-loving plant, requires a large quantity of water for maximum productivity. Banana production should be supported by an efficient irrigation system like drip irrigation. The plant should be irrigated immediately after planting. Sufficient water should be applied and field capacity also maintained. Excess irrigation will lead to root zone congestion due to the removal of air from soil pores, thereby affecting plant establishment and growth. And hence drip method is recommended for proper water management in Banana.

Application of Manure and Fertilizers for Banana Plants:

Banana requires a high amount of nutrients, which are often supplied only in part by the soil. The banana crop requires 7-8 Kg N, 0.7- 1.5 Kg P, and 17-20 Kg K per metric tonne yield. Banana responds well to the application of nutrients. Traditionally farmers use more urea and less phosphorous and potash.
In order to avoid loss of nutrients from conventional fertilizers i.e. loss of N through leaching, volatilization, evaporation, and loss of P and K by fixation in the soil, application of water-soluble or liquid fertilizers through drip irrigation (fertigation) is encouraged. A 25-30% increase in yield is observed using fertigation. Moreover, it saves labor and time and the distribution of nutrients is uniform.

Harvesting and Yield:

Harvesting of banana is done 10 to 12 months dwarf variety and 12 to 15 months for tall varieties. Signs of maturity of banana fruits are, fruit becomes plumpy and angles are filled in completely, when tapped it gives metallic sound, drying off of top leaves and change in color of fruits from deep green to light green.

Tall varieties like Poovan yield 15-25 tonnes/ha, while Dwarf Cavenshish yield 25-50 tonnes/ha. It can be stored at temperature slightly above $55°F$ and relative humidity of about 85-95% for about three weeks.

Banana Plantation

Pear Farming

Pear(Safou) is a brilliantly purple colored fleshy fruit. The botanical name for it is Dacryodes Edulis. It is also known as African plum, bush pear, purple pear, butter fruit or nsafu. It is also rich in oils, vitamins, and antioxidants. The average pear tree is able to produce notable quantity of fruits from the 3rd to 6th year of its age and can continue to do so until the 35th-50th year of its age.

Climate and soil requirement

Pear can be grown successfully on a variety of soils from sandy loam to clay loam provided it is deep, well drained, fertile, without the presence of hard pans of any type in the top 2 metres and does not have a pH value more than 8.7. In general, the pear tree thrives in cold and wet climate, where there is winter cold along with a cool summer.

When is best time to plant pear tree?

The planting of pear is done in winter while plants are still dormant. Pear requires full sun for best fruit set, fertile soil, well drained as well as good air circulation. Planting may preferably be completed within January. Generally one year old plants are used but 2 to 3 years old nursery trained plants can also be planted.

Weed Control

Light cultivation of the field should be done to manage the different kinds of weed flora. In pear orchards, 10 cm thick layer of paddy straw mulch (5.5 ton/acre) can be applied during second week of transplanting, mulching under the canopies of trees to suppress weeds.

Fertilizer Application

Nitrogen and Potassium are very important for leaf growth, flowering and fruit set, while Phosphorus is crucial for the development of a strong root system especially at the early stages of plant development. Calcium, Magnesium, Manganese, Zinc and Boron are also important for various stages of bud development, flowering and fruition and any deficiency will have a negative effect in fruit number, quality and general tree health. Mature fruit bearing pear trees need more fertilization than young trees that have not entered their fruition period. Checking the soil nutrients and pH is vital before applying any fertilization method.

Irrigation

The average young pear tree needs a lot of water in order to develop strong roots, leaves and finally fruits. It is beneficial to irrigate trees about once a week from spring to autumn (April – May to October). Soil texture and climate conditions determine the amount of water needed in order to harvest a fair yield. Irrigation should be restricted after fruit harvest. Bearing trees should be irrigated through flood method during summer months at an interval of 5 to 7 days so that fruit size may be increased. Over irrigation should also be avoided, as it will surely result in root rot. Drip, sprinklers or micro irrigation systems can be used for irrigation in pear orchards. Their main advantage is the protection of the tree buds from frost damage. Full coverage irrigation combined with the presence of cover crops often reduces the need for fertilizing, because the cover crop normally produces organic matter that is decomposed, releasing important nutrients to the soil.

Pests control

1. Hairy caterpillar (Euproctis sp.) are sporadic pests. Females lay eggs in clusters on ventral surface of leaves covered with yellow hair. On hatching, the young caterpillars feed gregariously on leaf lamina, skeletonzing the same completely.

2. Spider mites (Eutetranychus orientalis and Tetranychus urticae) attack during April-June. Initially yellowish-white specks appear on the leaves followed by leaf scorching and premature leaf fall. Infested leaves gather dust.

(i) Frequent irrigations during April-June.

3. Leaf-hoppers : Sometimes cause serious damage to the plants by sucking the cellsap from the leaves and tender fruits.

4. Bark eating caterpillar (Indarbela quadrinotata) : This pest causes damage by boring holes into the stem and branches and feeds on the bark under the cover of its excreta.

5. Fruit flies (Bactrocera dorsalis and Bactrocera zonata) are serious pests of pear fruits. Fruits nearing maturity are punctured by the fruit flies for egg laying and the maggots after hatching bore into the fruits thus render the fruits unfit for consumption.

(i) Harvest the ripening fruits and do not allow the ripe fruits on the tree.

(ii) Regular removal of fallen fruits from the ground and bury the infested fruits atleast at 60 cm depth.

Fruit Maturity and Harvesting

Harvesting of fruits should be done at proper stage of maturity. The fruits of Patharnakh mature 145 days after fruit set whereas Punjab Beauty take 135 days and Punjab Nectar and Punjab Gold take 140 days to reach maturity. The fruits should be picked by giving an upward twist to the fruit so that the spurs are not damaged.

Pear farming

Soybean Farming

The soybean or soya is a species of legume native to East Asia, widely grown for its edible bean, which has numerous uses. They are leguminous vegetable of the pea family that grows in tropical, subtropical, and temperate climates. They are rich in quality protein vitamins, minerals and digestible energy. They can be also turned into animal feed by baking the protein-rich fiber that remains after the oil is removed.

Soil requirement

The soil must be relatively smooth and friable before planting to allow good seed-to-soil contact. Planting populations based on the Soybean variety planted, but seeding rates usually range from 175,000 to 225,000 seeds per acre to provide quick in-row shading and weed management.

Soil fertility in organic systems is maintained through crop rotations, applications of seaweed and applications of manure, fish emulsion, or plant/animal-based products, such as feather meal. Soybeans fix nitrogen for the crop needs and can be grown without manure or compost unless phosphorus levels are deficient. Soil testing is suggested to determine the need for supplemental amendments. Subsequent crops must include rotations of grain crops and nitrogen-adding cover crops to maintain suitable fertility for future Soybean crops. Sample the soil in at least four places per acre to determine if lime is needed to adjust the pH value (6.5 to 7.0).

Seed selection

The first step to successful soybean planting and subsequent harvest is seed selection. Beans can be treated with any number of fungicide and pesticide products. Insecticide seed treatment may be desirable if the field is known for insect pests. Like any input, the cost must be weighed against yield gains.

Fungicide treatments are especially helpful if planting early in the spring, as wet, cool soil conditions can add to the risk of seedling disease. No-till fields will have cool soils longer into the season than tilled fields and commonly will have more seedling disease problems.

Seed Rate and Sowing in Organic Soybean Farming

Soybean has grown for grain purpose needs about 20 to 30 kg seed/ha but for fodder crop needs about 70 to 75 kg/ha during Kharif season and 100 to 120 kg/ha during spring. 45 to 60 cm X 2.5 cm spacing is good for the Kharif crop and 30 to 45 cm X 2.5 cm in the spring season.

The sowing must be done in lines 45 to 60 cm apart with the help of a seed drill or behind the plough. Plant to plant distance should be 4 to 5 cm. The depth of sowing must not be more than 3 to 4 cm under optimum moisture conditions. The seed rate of Soybean based upon germination percentage, seed size, and sowing time.

When is the best time to plant Soybean?

Planting of soybean requires adequate ground moisture and warm weather of at least 50°F. As this will ensure adequate soil temps is ideal.

For the upper Midwest, optimal planting time is generally April 25 to mid-May. By late May, producers will see a yield reduction.

Harvesting of Soybean

Soybeans are harvested in the fall when they're mature. As soybeans mature, their leaves start to turn yellow, and when the leaves of the soybeans turn brown and fall off, the soybean pods are left exposed and ready to be harvested. Harvesting soybeans involves a combine, or a large machine used to harvest grains. The combine's header harvests soybeans by cutting and collecting the soybean plants, removing their stems and pods and moving the harvested soybeans into its tank. Once the combine's tank is full, the harvested soybeans are emptied into a grain truck or wagon and taken to a grain dealer or storage facility.

Soybean

Scent leaf Farming

Ocimum gratissimum, popularly known as scent leaf is a fully developed flowering plant that is propagated by seed or cuttings. Scent Leaf is generally grown mainly in the tropical and sub – tropical regions of the world for its herb and spicy flavor. It is also used as a natural cure to many kinds of ailments and diseases.

Climate and soil requirement

It thrives best in a loamy soil with soil PH of 5-7. It requires a good amount of water during its sprouting process. When planted few weeks after the start of rainy season, it allows for fast growth and improvement of scent leaf plants.

Planting of Scent Leaf

Basically, there are two ways of planting these scent leaf – either by their seed and stem propagation.

Planting of Scent Leaf by Seeds

It can be done by taking the small seeds at the top of the fully grown scent leaf plant; these scent leaf seeds should be sprayed on the ground of the farm. In this way, they will grow, but it may take longer than the following method.

Planting of Scent Leaf by Stem

This is done by cutting the scent leaf stem of a fully grown scent leaf and inserts them in the ground of the farm. This is the most ideal way of propagating scent leaf for the best results. However, Scent leaf can be planted with other vegetables like waterleaf, spinach, eggplant. Etc

Pests & Disease Control in Scent Leaf Farm

Pests and fungi are the most common parasites causing plant disease. Most are microscopic (very small and can only be seen with the aid of a microscope) plants that feed on living green plants or on dead organic material. Fungi usually produce spores which when carried to a plant, can begin an infection. These spores may be carried from plant to plant by wind, water, insects and equipment. Fungi diseases in scent leaves are common during wet, humid seasons.

These diseases occur primarily on leaves, but some may also occur on stems and/or roots. Leaf diseases are the most common diseases of most plants. Various diseases that attack scent leaf includes: leaf spots, leaf blights, powdery mildew and downy mildew. They are usually controlled with fungicides, bactericides and resistant varieties.

Irrigation application

Drip irrigation is the best type of irrigation for the growing of Scent leaf. Watering of scent leaves should be done at least once a day for tender plants and twice a day for fully grown plants. Over irrigation and water lodging should be avoided as this may lead to poor growth and subsequently dead of the plant. Scent leave can be grown all year round if adequate irrigation is maintained.

Fertilizer/Manure

Scent leaves, just like other vegetables and most herbs does not require large amount of manure or Fertilizer to survive and thrive. Compost serve as the best manure and can be placed and allowed to decay to enrich the soil with nutrient. Animal dung can also be sprayed to make good manure too. Fertilizer; NPK can be applied 2weeks after sprout.

Weeding: Weed act as a hindrance to this plant, they compete with the plant for moisture, nutrition, light, and space. Effective weed control implies good control of weed throughout the growing season. You can control weeds by using chemicals or machines and perhaps, a combination of the two. The ultimate choice depends on the species of weeds involved, the level of infestation and the size of your budget.

Harvesting of scent leaf

Scent leaf, just like many other vegetable does not have a specified season of harvest. The plant continuously produces leaf even when they are harvested, till the plant gets weak and die off. They are harvested when the leaves grows large enough and turns dark green. Scents leaves take about 2 to 3 weeks to fully mature.

Scent Leaf

Beetroot Farming

The **beetroot** is the taproot portion of a beet plant, usually known in North America as beets while the vegetable is referred to as beetroot in British English, and also known as the table beet, garden beet, red beet, dinner beet or golden beet. It is one of several cultivated varieties of *Beta vulgaris* grown for their edible taproots and leaves (called **beet greens**). Beetroot is known for its medicinal purposes such as its laxative effects, cough remedy, headache medication, and even as an aphrodisiac.

Climate and soil requirement for beetroot cultivation

Spring and autumn are the best seasons for growing Beetroot, but it also grows well in summer and winter. Beetroot grows best under cool conditions and can be grown successfully almost all year round. Crops grown in cool weather produce superior quality roots (with high sugar content and dark internal color), though plant growth slows under prolonged cold conditions. Beetroot has a fair tolerance to moderate frosts at all growth stages. It grows well in warm weather but attains best color, texture, and quality in cool weather conditions. In hot weather the beetroot loses its color and quality. It causes the root to acquire alternating light and dark red concentric circles, which is referred to as zoning. The optimum temperature for germination of beetroot seeds is between 18 and 24°C; while soil temperatures between 4.5 and 30°C are acceptable.

Soil requirement

Beetroot farming yields the best results in loamy to sandy soils that are loose and well-drained. In clay soils, beetroot plant growth may be hampered and the shape of the beetroot may also be affected. Hence clay soil can be tolerated if the top has been loosened by the addition of lots of well-rotted organic matter

Heavy soils are not satisfactory for beets because the roots are likely to be unsymmetrical in shape when grown on such soils. A pH range between 6.3 and 7.5 is most appropriate for cultivation of beetroot. Beetroot likes neutral, moist, fertile soil without too much lime or acidity.

As acidic soils may cause nutrient deficiencies in Beetroot, care should be taken to ensure pH levels in the soil are not high. A mature beetroot plant is fairly tolerant of salt in the soil; however, beetroot seedlings are comparatively sensitive to the salinity of the soil.

Propagation of Beetroot:

Propagation in beetroot farming process is done mainly through seeds. Seeds of beetroot can be sown about 1.5 cm deep leaving a distance of about 7cm in between. The space between the rows must be 30 to 40 cm. Mid-April- Mid July is the ideal time for sowing though it may vary with climatic conditions of different regions. Beetroot should be grown quickly if growth is checked; the edible roots become tough and indigestible. Watering should be constant but not over the top. In dry conditions thorough watering is necessary. The seedlings will appear in about 15 days, depending on the weather. When the seedlings are about 3 to 5 cm high, thin them to a spacing of 7 to 10 cm to out so that there is only one seedling in one place. Remove the weakest seedlings, leaving only the strongest for growth. The beetroot takes about 2 months to grow from sowing to maturity. The mature height of the plant is about 9 cm.

Seedlings of Sugar Beetroot

Garden beet is primarily a cool season crop but grows well in warm weather and hence can be grown during winter all over the plains. Generally, sowing of the beet can be done any time from August to November. In some parts like Bangalore, it is sown during June-July also. Like other root crops, it is also preferred to be sown on ridges by keeping the row to row distance 30 to 40 cm and plant to plant about 15 -25 cm. It requires about 10 -14 kg of seed to sow one-hectare of area. The seed is sown 2 -3 cm deep in the soil and irrigated immediately after sowing in light soils.

At some places, the seeds are soaked for about 12 hours before sowing to facilitate better germination in the field. The seedlings come up in bunches since each seed sown is really a fruit containing 2-6 seeds. Only one plant should be allowed to grow in each clump.

Seed Treatment for beetroot cultivation

It is recommended to treat beetroot seeds before planting to improve their germination. The high nitrate content in dry seeds, as well as the ammonia produced by the bacteria, suppresses the germination of Beetroot seeds. Therefore, the seeds should be washed in running water for at least an hour before planting.

After washing, the seeds should be soaked for 20 minutes in a 0.5 % Aretan solution, and they should be allowed to dry for at least 6 hours at room temperature before sowing.

Fertilizer requirement of Beetroot:

It has been reported that one tonne of beetroot removes 2 kg of nitrogen, 4.5 kg of phosphorus and 4.5 kg of potash. Usually, 60-70 kg of nitrogen, 100-120 kg of potash per hectare is applied to the soil at the time of field preparation. In addition, 10-15 tonnes of farmyard manure (FYM) may be mixed up thoroughly in the soil at the time of field preparation. The beet crop is also susceptible to boron deficiency causing an internal break down of the roots. Hence borax or boric acid may be applied in boron deficient soils.

Water requirement for growing Beetroot:

All with most of the vegetable crops, good irrigation conditions are also necessary for good yield and quality in beetroot farming. Beetroot seeds require plenty of water to start the germination process. The roots will take moisture from the soil once they're established. During the growth period, 300-350 millimeters of water is needed for the entire growth period of the crop, i.e. almost about 4 millimeters per day. Excess-watering should be avoided. This causes beetroot to produce more leaves and less root, risking them "bolting" (flowering and not producing a vegetable). Under-watering creates woody roots. Also once they have sprouted; it should be watered every 10-14 days in dry spells. Other than when the weather is unnaturally dry, normal rainfall should be fine. Irrigation depends on soil type.

Pests and diseases of Beetroot:

Beetroot is relatively free from pests and diseases. In Beetroot Farming, leaf spot is the most common disease in Beetroot Farming, causing a conspicuous spotting which is very easily recognized. The disease is favored by warm, wet conditions. Rotating crops, allowing 2 to 3 years between planting beetroot in the same garden bed, will help to suppress the leaf spot. Root-knot nematodes cause galls or swellings on the roots. Heavily infested plants are usually stunted, with the main root malformed. Removing the affected leaf parts as soon as the damage is seen, burning them, isolating the affected plant and treating lightly with diluted insecticides and fungicides goes a long way in checking the damaging effects of diseases and pests. Contact your nearest agriculture research department for pest and disease control in Beetroot Farming.

Harvesting and storing of Beetroot:

Harvesting of beetroot can begin around nine weeks after sowing the seed. At this stage, the bulbs will be about 2.5cm (1 inch) in diameter and they will be at their most tender – best for salads. These first pickings should be evenly applied over the growing area to give the remaining beetroot good room to grow larger. Continue to harvest as required until the beetroot reach about 8 cm (3 inches) in diameter. At this point, it is best to harvest all the beetroot and store them. If they are left in the ground much longer, they will become woody and not taste so good. Another sign that the roots are ready for harvest is when the foliage starts to go limp. Beets are harvested as they attain a diameter of 3-5 cm. They are usually pulled by hands; the tops are removed and after washing the roots are graded according to size. In advanced countries, the mature roots are mechanically harvested; the topped, washed, graded and finally packed in polythene bags. Removal of tops and packing in polythene bags lengthens the shelf-life of beets by reducing water loss during transit and storage. Small sized bunched beets are also in great demand in some countries. Hence, after harvesting by hand the dead and injured leaves are removed and then tied in bunches of 4-6 beets with their tops on. Oversized beets are not in demand, because they are tough and woody and cracks appear on the surface.

Weed Management in beetroot cultivation

Weeds are detrimental for the beetroot crops especially at the early stages of the season. Crop yield can be severely affected by weed problems. Weeding can be done manually or with the use of weed killers or herbicides depending on the size of the farm.

Beetroot

MANAGEMENT OF OTHER CROP & HORTICULTURAL PESTS

I. Squirrel

The squirrel (*Funambulus pennenti*) lives in nest of twigs and leaves on trees. Fruits and seeds are its major food. In orchard, it causes damage to a variety of fruits such as guava, ber, peach, plum, pomegranate, mulberry, grapes and jamun. For its control, traps should be used. Keep the traps on squirrels active sites during morning and evening as these are more active during these periods. Methods for the use of traps are the same as given for rats and mice.

II. Rats and Mice

Rats and mice usually live in burrows on the ground, possess acute senses of smell and taste, and are very selective in food choice. They are prolific breeders, extremely adaptable and intelligent pests and thus their control poses difficulties. Out of 8 species of rodents in fields, the lesser bandicoot rat, *Bandicota bengalensis* is most predominant under irrigated conditions and Indian Gerbil *Tatera indica* in dry and sandy soils.

The rats and mice attack seeds and seedlings of vegetable crops at growth stage and fruits at ripening stages. Seedlings are also destroyed under heaps of soil made by rats. The lesser bandicoot rat during burrowing, loosens the soil resulting in the drying of plants.

Methods of Control

The performance of different control methods vary in different situations and at different stages of the crop. Therefore, best control can only be achieved if these methods are adopted properly at appropriate timings.

A. Mechanical Control

i) During irrigation of vacant harvested fields, rats coming out of flooded burrows should be killed with sticks.

ii) Traps can be used to control rodents. Place 16 traps per acre at damage and activity sites of rodents. Kill the trapped rats by drowning in water and the interval between two trapping at the same location should not be less than 30 days.

B. Environmental Control

Weeds, grasses and bushes should be removed as these provide shelter and food to rodents. Highly infested bunds, water channels and field pavements should be periodically rebuilt to destroy permanent rat burrows. Keep the height and width of bunds to the minimum.

Waste lands along roads, canals, railway lines, other uncultivated areas and forest strip serve as hiding places for rodents. So, to protect the adjoining crops, rat control operations must be carried out in these areas also.

C. Biological Control

Owls, kites, eagles, falcons, cats, mongoose, jackals, snakes and lizards are the natural predators for rats and mice. These should be protected.

D. Chemical Control

Baiting Technique

Poison bait preparations: The acceptance of poison baits by rodents depends upon the quality, texture, taste, odour etc. of the baiting materials. Therefore, the recommended baiting materials should be used for preparation of poison baits.

(i) 2% Zinc phosphide bait: Smear 1 kg of bajra or sorghum or cracked wheat or their mixture with 20 g of edible refined oil and mix it thoroughly with 25 g of 80% zinc phosphide powder.

Caution: Never add water in zinc phosphide bait and always use freshly prepared bait.

(ii) 0.005% Bromadiolone bait: Mix 20 g of 0.25% bromadiolone powder, 20 g of edible refined oil and 20 g of powdered sugar in 1 kg of any cereal flour or bajra, sorghum or cracked wheat.

Poison Bait Placement and Timings

Burrow baiting: Rat burrows can be easily located in the fields, on bunds, water channels and surrounding waste lands. Close all the burrows in the evening and in the re-opened burrows on the next day, insert a paper boat containing about 10 g of poison bait about 6 inches deep in each burrow. In case of burrows of the lesser bandicoot rat, gently remove the fresh soil from the burrow opening to locate the tunnel and then put the poison bait deep inside it.

Crop baiting : Place about 10g of zinc phosphate or bromodiolone bait at 40 bait points per acre on dry sites and inside the crop throughout the field covering runways and activity sites of rats.

Pre-baiting : To increase the efficacy of zinc phosphide bait do pre baiting. Place bajra & sorghum or cracked wheat or their mixture smeared with oil on pieces of paper, 10g each at 40 bait points per acre for 2-3 days.

Safety Measures

Since the rodenticides are very toxic to humans, domestic animals, pets and birds, the following safety measures must be adopted.

1. Keep the rodenticides and poison baits away from the reach of children, domestic animals, pets and birds.

2. Mixing of rodenticide in the baiting material should be done with a stick, spade or by wearing gloves. Avoid the contact of poison with mouth. Wash exposed skin and hands after mixing.

3. House hold utensils should never be used for preparation of the rodenticide baits.

4. Polythene bags used for storage and carrying the rodenticide bait should be buried after use.

5. Collect and bury left over rodenticide bait and dead rats from the orchard.

6. Zinc phosphide is toxic and there is no antidote for it. In case of its accidental ingestion, induce vomitting by inserting fingers in the throat and rush to doctor. Vitamin K is antidote for bromadiolone, it can be given to the patient under medical advice.

Integrated approach

No single method is 100% effective in controlling rats and mice. Left over population reproduce reaching the original size in short time. Therefore, adopt an integrated approach by using different methods at different stages of the crop.

Village level campaign

Control of rats and mice in smaller area usually become ineffective due to their migration from the surrounding untreated fields. Therefore, for better results, village level anti-rat campaign to cover maximum possible area, both cultivated and uncultivated, should be organized.

III. Birds

Birds, in general, are both useful and harmful to fruit crops. Even the same species may be beneficial or problematic in different situations. Only a few of about 304 species of birds of Punjab cause problems in orchards. The rose-ringed parakeet is the only bird that seems to be exclusively harmful to farmer's interests.

Harmful Birds

Several fruits are damaged by birds at the bud stage and ripening stage. Parakeet is the major bird pest causing serious damage to guava, peach, pear, grapes, mango and ber. House crows damage peach, plum and grapes. The major damage to grape is caused by mynas, especially the bank mynas.

MANAGEMENT OF BIRD DAMAGE

A. Mechanical Control

1. Make false gun-shots at different intervals to scare the birds.

2. Frequent beating of drums and use of Gopia at different points in the orchard is very effective against the birds.

3. Covering the vines of grapes and isolated fruit trees with nylon nets prevents the bird damage.

4. Fixing of scare crows i.e. a discarded earthen pot painted to stimulate human like head supported with wooden sticks and clothed in human dress to give a human like appearance is one of the most effective traditional techniques to keep the birds away. Position, direction and the dress of the scare crow should be changed at 10 days interval. The height of the scare crow should be 1 metre above the plant height.

5. Use automatic bird scarer by shifting their position periodically and supplementing their noise with actual gunfire's. The other simplest method is the use of rope-crackers. It involves tying of sets of small fire crackers at a distance of 6-8 inches apart on a rope and igniting it from the lower end. The explosions caused by fire crackers on catching fire at different intervals scare the birds feeding on fruits. Fix the rope-crackers in the centre of the orchard.

B. Cultural Practices

1. As far as possible, sowing of maize and sunflower crops should be avoided in and around the orchards. Prepare the nurseries of fruit trees at the right and recommended time and keep it covered with thorny bushes or nylon net.

2. The trees in the vicinity of fruit nursery should be pruned so that birds may not establish their nests on them.

3. Weed and grasses should be removed regularly as they provide shelter to birds.

4. Birds should be prevented from nest making on trees. Older nests on trees should be removed before the birds start their breeding activities.

5. Fruit trees planted solitary should be covered with nylon net with mesh size of 2.5 to 5.8 cm. Grapevines should be covered with 1.25 cm size nylon net. It should completely cover the vines from above and also touch the ground on all sides. This nylon net is effective in preventing the entry of even small birds like sparrow. Wasps and other flying insects which cross the net can also fall prey to birds sitting outside the netted area.

6. Grow the less costly crops like daincha or millet around the orchard, these crops are preferred by the birds and even prevent bird damage to orchard. In addition to this, these crops being tall also act as wind breakers and help in preventing lodging of small plants during stormy/rainy days.161

7. Preferably, fruit trees should be planted away from the perching sites of birds or cluster of trees and also no electricity wires should pass above the fruit trees.

8. To prevent the parakeet damage, fruit trees should be planted in large block area (at least two-three acres) because parakeets avoid feed/venturing in the interior of the orchard.

C. Use of Reflective Ribbon

Reflective ribbon should be tied with wooden stick or bamboo at least 1-3 feet above the fruit nursery or at the outer side of the orchard. If there is more than 10 meters gap between wooden sticks, then additional wooden stick should be installed for support. Ribbon should be installed from north to south direction. Ribbon should be slightly twisted and loosely tied. By using the ribbon in this way, the rays of rising sun from east and setting sun from west fall on the ribbon, due to which the ribbon reflects the light and also produces wavering voice with the blowing wind, which scares the birds away.

D. Alarming Calls

Playing of cassettes (available at Centre for Communication and International Linkages, PAU) of distress or flock calls of parakeets and crows respectively in a tape-recorder at peak volume for 1/2 hr twice each in the morning between 7.00 to 9.00 a.m. and in the evening at 5.00 to 7.00 p.m. respectively, with a pause of 1 hour, scare the birds or halt their activities in orchards. Use of distress or flock calls remain effective for 15-20 days. Better results can be obtained by using this technique in sequence or in combination with other methods as an integrated pest management. For covering larger area, use of amplifier or additional speakers (as per requirements) can be done.

Conservation of Useful Birds

Predatory birds like owls, falcons, hawks, eagles, kites, etc. eat a large number of rats and mice. A single owl normally eats 4-5 rats a day. Insect eating birds like drongos, babblers, shrikes, lapwings, mynas, and many other small birds like sparrows and weaver birds feed a large number of insects to their young. A single pair of house sparrows feeds insects to their young about 250 times a day. Therefore, the useful birds should not be killed.

APPENDIX - I

Fertilizer Sources for the Supply of Nitrogen, Phosphorus and Potassium

(A) Nutrient contents of different fertilizers

Fertilizer	N (%)	P_2O_5 (%)	K_2O (%)	Other
Ammonium sulphate	20.5	-		-
Ammonium chloride	25.0	-		-
Calcium ammonium nitrate	25.0	-		-
Urea	46.0	-		-
Superphosphate (single)	-	16.0		-
Diammonium phosphate	18.0	46.0		-
Urea-ammonium phosphate	28.0	28.0	-	-
Nitro phosphate	20.0	20.0	-	-
Sulphate of potash	-	-	48.0	
Sulphated phosphate	13.0	33.0	-	15(s)
Muriate of potash	-	-	60.0	

Manganese Sulphate	-	-	-	30 (Mn)
Zinc Sulphate (Heptahydrate)	-	-	-	21 (Zn)
Zinc Sulphate (Monohydrate)	-	-	33 (Zn)	
Ferrous Sulphate 7 H2O	-	-	-	19 (Fe)
Copper Sulphate S1 H2O	-	-	-	24 (Cu)
Gypsum		-		

(B) Quantity of the fertilizer to give 1 kg of nutrient

For 1 kg of N	
Calcium ammonium nitrate	4 kg
Ammonium chloride	4 kg
Ammonium sulphate	5 kg
Urea	2.2 kg
For 1 kg of P2O5	
Superphosphate	6.2 kg
Diammonium phosphate	2.2 kg
Urea-ammonium phosphate	3.6 kg
For 1 kg of K2O5	
Muriate of potash	1.7 kg

Note: Urea-ammonium phosphate (28-28), and diammonium phosphate (18-46) contain both nitrogen and phosphorus. By adding one kg of phosphorus (P2O5) through these fertilizers, one kg nitrogen (N) from urea ammonium phosphate and 400 g of N from ammonium phosphate is also added. This point must be taken into account while using two fertilizers.

APPENDIX - II

Waiting Periods of Different Pesticides in Fruit and Vegetables Crops

Recommended Pesticide	Crop	Waiting Period (Days)
Malathion	Brinjal, Okra, Cabbage, Grapes	1
Quinalfos	Brinjal, Cabbage (Autumn)	4
	Cabbage, Cauliflower (Winter)	7
	Kinnow	7
	Grapes	10
Chlorpyrifos	Peas, Chilli	7
Cypermethrin	Tomato, Cabbage, Okra	1
Fenvalerate	Cabbage, Brinjal, Cauliflower	1
Deltamethrin	Kinnow	1
Triazophos	Okra, Kinnow, Cauliflower	7
	Brinjal	4
Dicofol	Brinjal, Cucumber	1
Ethion	Brinjal	5
	Cucumber	7
	Pear	1
Fenvalerate	Pear, peach, Guava	2
Spinosad	Cauliflower	7
	Cabbage	5

Indoxacarb	Cauliflower, Cabbage	3
Propargite	Brinjal	1
Cypermethrin	Kinnow	3
Carbaryl	Grapes	3
Profenofos	Tomato	5
Flubendiamide	Tomato	3
Emamectin Benzoate	Okra, Cabbage Cauliflower	3

APPENDIX - III

General Recommendations Regarding Safe Use of Pesticides

1. Read the label carefully and follow the manufacturer's instructions.
2. Keep pesticides in labelled containers only.
3. Store pesticides in a safe and locked place, out of reach of children, irresponsible persons and pets.
4. Never store pesticides near foodstuffs or medicines.
5. While handling the dangerous pesticides, the necessary protective clothing and devices must be used.
6. Do not tear open the pesticides bags, but cut them with a knife.
7. The preparations of spray solutions from concentrated dangerous pesticides should be done in drums using long sticks to protect the operator from splashing and to permit stirring from a standing position.
8. Wash hands thoroughly with soda and water (i) every time the sprayer/duster is filled with pesticides, (ii) before eating, drinking or smoking and (iii) at the end of the day's work.
9. Water contaminated, as a result of washing the equipment and drums, must be disposed off by scattering it over barren land.
10. Do not blow, suck or apply your mouth to any sprinkler, nozzle or other spraying equipment.
11. Operators should not work for more than 8 hours a day. Those engaged in handling dangerous pesticides should be checked up by a physician periodically.
12. Separate working clothes should be used. They should be washed and changed as frequently as possible.
13. Do not use the empty containers of pesticides for any purpose. Destroy them by making holes and burry them afterwards.
14. Do not burn pesticide cartons, but bury them deep.
15. The worker should not smoke, chew, eat or drink in the spraying area or while spraying
16. A worker suffering from cold or cough should not be engaged for spraying.
17. Spray should always be done in direction of the blowing wind to avoid skin exposure and inhalation.

First Aid Precautions

In case of pesticide poisoning, call a physician immediately. Awaiting the physician's arrival, apply the FIRST AID.

1. Swallowed Poisons

a) Remove poison from the patient's stomach immediately by inducing vomiting. Give one teaspoonful (15 g) common salt in a glass of warm water (emetic) and repeat until the vomit fluid is clear. Gentle stroking or touching the throat with a finger or placing the blunt end of a spoon will help induce vomiting when the stomach is full of fluid.

b) If the patient is already vomiting, do not give common salt in warm water and follow the specific directions as suggested.

2. Inhaled Poisons

a) Carry the patient (do not let him walk) to fresh air immediately.

b) Open all the doors and windows.

c) Loosen all tight clothing.

d) Apply artificial respiration if breathing has stopped or is irregular. Avoid a vigorous application of pressure to the chest.

e) Cover the patient with a blanket.

f) Keep the patient as quiet as possible.

g) If the patient is convulsing, keep him in bed in some dark room.

h) Avoid any jarring noise.

3. Skin Contamination

a) Drench the skin with water (giving a shower with a hose or pump).

b) Apply a stream of water to the skin while removing the clothing.

c) Clean the skin thoroughly with water.

d) Rapid washing is most important for reducing the extent of injury.

4. Prevention of Collapse

a) Cover the patient with a light blanket.

b) Do not use a hot-water bottle.

c) Raise the feet of the patient on the bed.

d) Apply elastic bands to arms and legs.

e) Give strong tea or coffee to drink.

f) Give hypodermic injection of stimultants, such as caffine and epinephrine.

g) Give fluid administration of dextorse 5% intervenously.

h) Give blood or plasma transfusion.

i) Do not exhaust the patient by too much or too vigorous treatment.

5. Eye Contamination

a) Hold eyelids open.

b) Wash the eyes gently with stream of running water immediately. A delay of even a few seconds greatly increases the extent of injury.

c) Continue washing until the physician arrives.

d) Do not use chemicals. They may increase the injury.

APPENDIX - IV

Some Other First Aid Measures

1. Cut Injury

a) The first aid treatment of cut injury depends upon the date and extent of injury.

b) In first aid, one should clean the wound with antiseptic lotion.

c) If it is bleeding profusely, tight bandage without ointment is to be given.

d) The injured part should be kept raised or elevated.

e) If there is any associated fracture, a proper split or support should be given. But the patient should be brought to the hospital at the earliest possible.

2. Snake Bite – Preventions

a) In snake infested regions, long trousers, high shoes or legging and gloves should be worn. Most important is to look where to step while walking.

First Aid

a) Re-assure the complete rest to the victim to retard the absorption of venom. A wide tournaquet of cloth should be a few tied centimeters proxymal or above the site of bite. It should be tight to an extent that a finger passes below it with difficulty.

Suction of venom should be done by giving 1 cm linear and 1/2 cm deep incision at the mark of the fangs after applying an antiseptic lotion. Suction should preferably be done with rubber bulb, breast pump or with mouth after ensuring that there is no oral lesion. It should continue for about an hour. If done promptly, 50% of the venom can be removed.

3. Electric Injuries – Preventions

a) Educate the electric hazards to everybody.

b) Proper installation of electric appliances, grounding of telephone lines, radio and television areals, use of rubber gloves and dry shoes when working with electric circuit.

First Aid

a) Prompt switching of the current, if possible.

b) Immediate removal of the victim from the contact with the current without directly touching him. Rescuer should use a rubber sheet, a leather belt, a wooden pole or any other non-conductive material to detach him.

c) If the victim is not breathing, mouth respiration should be given.

d) If no pulse is felt, cardiac massage (pressure on left side chest) should be given.

e) In mild cases, local treatment of burnt part is required.

4. Honey bee bites

a) Cooling of the part with ice pads.

b) Removal of stings.

c) Cleaning with soap and water.

d) Local and systemic anti allergics to be given.

e) Perfumes and bright colours attract these insects and should be avoided.

f) Sensitive person can have severe anaphyllatic shock with even a single bite. Every such patient must get the medical aid from a doctor.

REEFRENCE

(1) Farm Chemicals Handbook, 1994

(2) Health hazards of Pesticides and its management (1996) Voluntary Health Association of India

(3) Essentials of Forensic Medicine and Toxicology (1999) by Narayan Reddy

(4) National Poison Information Centre, AIIMS, New Delhi

(5) Ministère de l'Agriculture des Ressources Naturelles et du Développement Rural (MARNDR).-

(6) Manuel Pratique de Conservation de Sols d'Haiti.- Haiti, 1999

(7) Ministère de l'Agriculture des Ressources Naturelles et du Développement Rural (MARNDR).-

(8) Recensement General de l'Agriculture (RGA), Resultats Provisoires Departement du Nord-Est.- Port-au-Prince, 2008/2009

(9) SoCo Project Team.- Final report on the project 'Sustainable Agriculture and Soil Conservation (SoCo)', 2009

(10) http://www.fao.org/ag/fr/magazine/fao0gapfr.pdf

(11) http://www.ma.auf.org/erosion/chapitre1/VI.Lutte.html

12) https://www.soyconnection.com/soy-farms/soybean-farming

www.ingramcontent.com/pod-product-compliance
Lightning Source LLC
Chambersburg PA
CBHW040323220526
45473CB00009B/2550